# CONTENTS

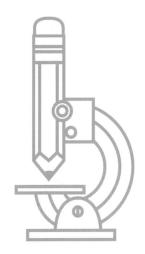

# HOLE IN ONE

The central core of our galaxy, the Milky Way, is a supermassive black hole. The black hole is commonly referred to as Sagittarius A* – it contains the mass of an estimated 4.3 million Suns.

Nothing can escape the gravity of a black hole – including light. This means that scientists have to be clever when studying them. As they cannot 'see' the black holes usually, they use other indications like radiation emissions to know that they are there.

# ODD SCIENCE
# SPECTACULAR SPACE

James Olstein

PAVILION

Space is fascinating. From the planets in our galaxy to mysterious black holes, space lightning to runaway stars, moonquakes to ultraviolet radiation.

Deepest space has all sorts of strange secrets – like the presence of toxic grease floating through the Milky Way, exoplanets that are made of pure diamond and the true colour of the Sun.

This book will reveal things you've never heard before. It will tell you unknown facts about space itself, and it will surprise you with weird and wonderful information about life in a space station.

This book will tell you what can happen to astronauts' feet, the problem with itchy noses on a spacewalk and how astronauts train on Earth for their big launch into outer space.

Quirky, strange and cool: come inside the world of odd science.

_____

*For my family.*

## ONE OF A KIND

Our solar system is very unusual.
Most other sun-like stars studied
to date do not have large planets
orbiting them.

# GETTING ENOUGH SPACE

Spacecraft sent from Earth have now visited
all the planets in the Solar System.

## MYSTERIOUS MERCURY

Only two spacecraft have ever got close to our smallest planet. They were used to map its cratered surface and study its atmosphere.

## HOT STUFF

Venus is a turbulent, hot planet. There have been many failed attempts to visit it, but more than 20 missions have been successful.

## MIGHTY MARS

Scientists have sent many missions to Mars, but only 18 have been fully accomplished. Eight rovers have landed on the surface – four still live there.

## NINE IS FINE

Nine spacecraft have studied Jupiter. The first to make it, in 1973, was NASA's Pioneer 10, flying 200,000 km above the clouds that cover the gas giant.

## RING ME

Four spacecraft have visited Saturn. As well as observing the planet's ring system, they also studied its many moons.

## ONE AND ONLY

So far, only one spacecraft has reached distant Uranus. All of the information we have comes from this one visit and the use of powerful, ground-based telescopes.

## FAR FAR AWAY

From Earth, Neptune is not visible to the naked eye. It's the furthest official planet from the Sun and has also only ever been visited once.

## CAST IRON

The surface of Mercury looks a lot like our Moon, but its core has more iron than any other planet or moon in the Solar System.

## HOT HOT HOT

Venus' thick atmosphere traps heat, creating a greenhouse effect. The surface temperature of the planet can bubble up to 465 °C, hot enough to melt lead!

## STORMY WEATHER?

On Mars, dust storms can engulf the entire planet and then rage for weeks on end.

## SOLID CENTRE

Although Jupiter is mostly gas, at its centre there is a core made of ice, rock and metals.

## FLATLAND

Saturn is gassy and rotates at high speed, helping to make it the flattest planet in the Solar System.

## ICY RECEPTION

Uranus is known as an 'ice giant' – it has the coldest atmosphere of any of the planets.

## BRIGHT AND BREEZY

Raging gales on Neptune can reach speeds of 2,400 kph – the fastest planetary winds measured so far in the Solar System.

## SURFACE TO AIR

Unlike Earth, the gas giants – Jupiter, Saturn, Uranus and Neptune – have no solid surface for anyone to walk on!

## TIME TABLE

A day is the time a planet takes to rotate once. A year is measured as one rotation around the Sun. Each planet is a different size and distance from the Sun, so no two have the same length of day or year.

## TIME WARP

A day on the surface of Mercury lasts 176 Earth days, but a year on the planet only takes 88 Earth days.

## UPSIDE DOWN

On planet Venus, a day lasts for 117 Earth days and yet a year's orbit takes 243 Earth days.

## TRAVEL TIME

Light from our Sun takes over eight minutes to reach us here on Earth.

## LIFE ON MARS

At 24 hours and 39 minutes, a Martian day is very close to our own, but a year lasts for 687 Earth days.

## BIG YEAR

One year on Neptune lasts for a staggering 165 Earth years! Its days are only 16 hours long.

## SLOW MOVER

A day on Uranus lasts for around 17 hours. It circles the Sun super-slowly, taking 84 Earth years to complete each annual orbit.

## RING IN THE YEAR

Saturn spins quickly, completing a day in 10 hours and 33 minutes. A year is equal to around 29.5 Earth years.

## HALF DAY

Jupiter rotates around the Sun once every 11.8 Earth years, but it has the shortest day of all – 9 hours and 55 minutes.

## OLD FRIEND

NASA has estimated that our Sun is 4.5 billion years old, around halfway through its life.

## ONE IN A MILLION

Our Sun is considered an average-sized star. One million Earths could fit inside it.

# SUN SIZE

The Sun accounts for 99% of all of the mass in our Solar System. The star is a huge ball of hydrogen gas held together by gravity. In around 5 billion years the Sun will have consumed all its hydrogen. It will grow larger and larger, in what is called a red giant phase, until it engulfs Mercury, Venus and Earth. After this phase, the Sun will collapse – into a dense white dwarf the same size as Earth today.

## COLOUR-FULL

To our eyes, the Sun appears to be white. However, it is actually a blend of all the colours mixed together. Rainbows are light from the Sun, separated into all of its colours.

# SUNNY HOME

We all live inside the atmosphere of the Sun! The star's corona, or atmosphere, reaches as far out as Neptune and beyond. Earth is protected from the Sun's heat by our own atmosphere and powerful magnetic field.

# TOO HOT TO HANDLE

At its centre, the Sun reaches searing temperatures of 15 million °C.

## JUST PASSING BY

Mercury in transit occurs when the planet passes in front of the Sun at an angle that makes it visible to astronomers on Earth. It only happens 13 times in a century.

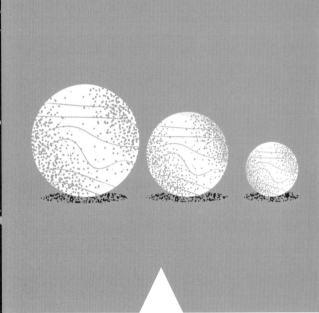

## SHRINKY DINK

Mercury is slowly cooling down, a process that causes the planet to shrink every year.

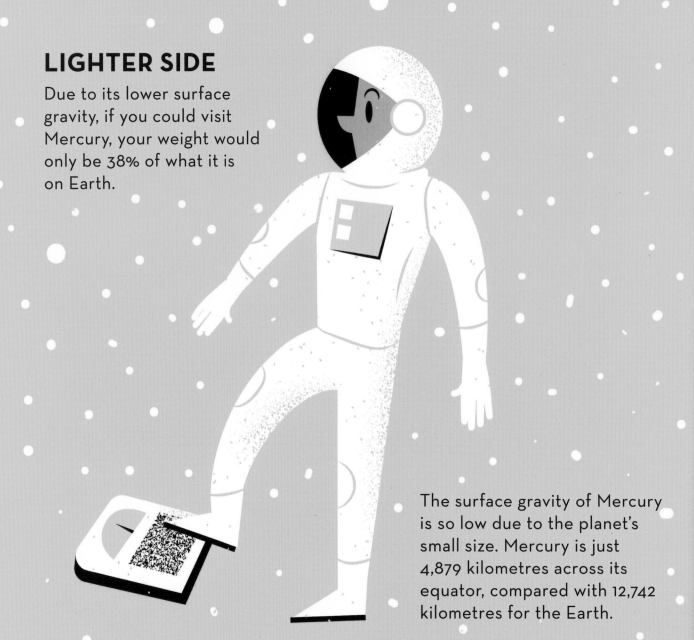

## LIGHTER SIDE

Due to its lower surface gravity, if you could visit Mercury, your weight would only be 38% of what it is on Earth.

The surface gravity of Mercury is so low due to the planet's small size. Mercury is just 4,879 kilometres across its equator, compared with 12,742 kilometres for the Earth.

## IRON IT OUT

As Mercury's iron core contracts, it causes the planet's rocky crust to form wrinkles and mountain-like ridges on the surface. These are called lobate scarps, and they can reach up to 2km high.

## MIRROR MAGIC

The third brightest object in the night sky, after the Sun and Moon, is Venus. This is due to light reflecting off the clouds in its atmosphere.

MERCURY

VENUS

EARTH

MARS

JUPITER

SATURN

URANUS

NEPTUNE

## GO YOUR OWN WAY

Venus and Uranus rotate in the opposite direction to the other planets in the Solar System. Because of this phenomenon, if you were visiting Venus you would see the Sun rising in the West and setting in the East.

## TROPICAL PARADISE

After radio-mapping Venus' surface, scientists believe that the planet did once have oceans. These seas evaporated when the temperature warmed up.

## DRAWN TO IT

Earth has a magnetic field that protects us from solar winds and other forms of harsh space radiation.

## TAKING IT EASY

Due to tidal effects and other factors, the Earth's rotation is gradually slowing down. Over time, this will make a day last very slightly longer.

## MARS ROCKS!

On December 27, 1984, US meteorite hunters found rock from the planet Mars in Allan Hills, Antarctica. The most popular theory remains that this rock was blasted from the surface of Mars by a meteor impact about 4 billion years ago, before eventually falling to Earth.

As of 2019, 214 meteorites found on Earth have been identified as Martian.

## GET IN SHAPE

The Earth is not actually a perfect sphere – it only appears round when viewed from space. The 'blue marble' has a bulging middle and is slightly flatter at each of its poles.

## BELT UP

The Van Allen belts are two rings of radiation that surround Earth. The outer belt is made up of billions of particles from the Sun trapped within the Earth's magnetic field. The inner belt is created by cosmic rays colliding in our upper atmosphere.

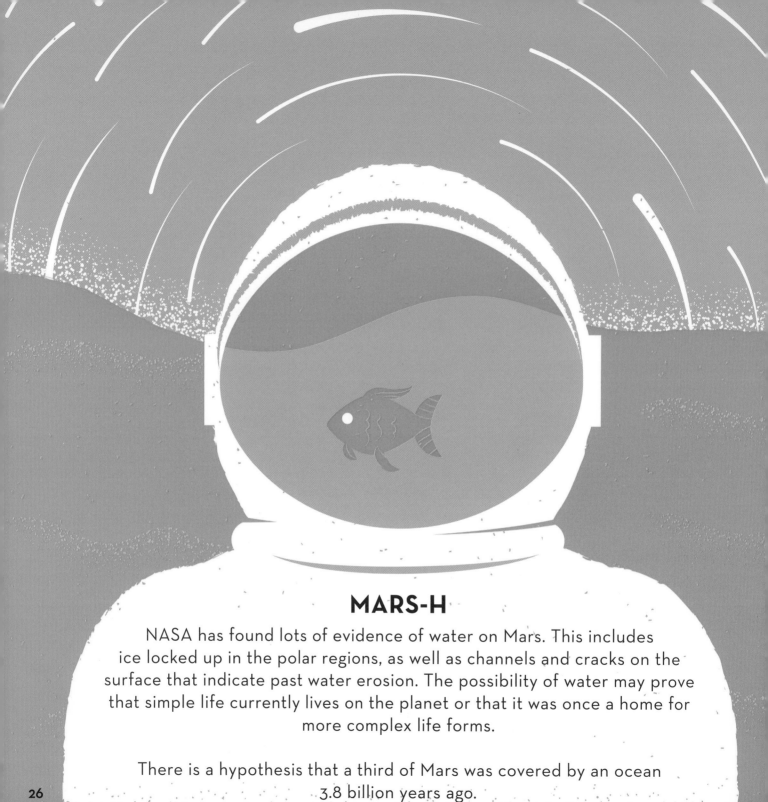

## MARS-H

NASA has found lots of evidence of water on Mars. This includes ice locked up in the polar regions, as well as channels and cracks on the surface that indicate past water erosion. The possibility of water may prove that simple life currently lives on the planet or that it was once a home for more complex life forms.

There is a hypothesis that a third of Mars was covered by an ocean 3.8 billion years ago.

## GOT THE BLUES

On Mars, the sky appears red during the day, but as night falls the sunset turns blue. This is caused by fine dust particles forming a blue halo around the Sun on Mars – which is only seen when light passes through the halo while coming over the horizon.

27

## ATMOS-FEAR

Scientists believe that Mars' atmosphere was once similar to ours, before it got stripped away by blistering solar winds and ultraviolet radiation. Today the weak atmosphere of Mars is composed of 96% carbon dioxide – Earth's is 78% nitrogen.

## KING OF THE RING

One day, Mars may have its own ring. In millions of years' time, its innermost moon, Phobos, is likely to either crash into the planet or break apart, creating a ring of rocky debris.

# CRAZY CLOUDS

Jupiter's swirling features are incredible. Bands of clouds cross above it in shades of white, brown and orange. At the deepest levels, there are blue clouds, too.

## SPOT ON

The Great Red Spot on Jupiter's surface is actually an immense hurricane. At its largest documented size, the churning storm is as wide as three planet Earths. It is thought that the Great Red Spot has been raging for at least 350 years.

## FLOAT ON

If you could shrink Saturn to fit into a vast body of liquid it would float! It is the only planet in the Solar System with a density that is less than water. Saturn's density is 0.687 grams per cubic centimetre. Earth is the densest planet in the solar system at 5.51 grams per cubic centimetre.

## BAND TOGETHER

Saturn's atmosphere is made up of amazing layers of clouds. The band in the upper level is made of a special combination of ammonia and ice.

# RING AROUND

Although it is hard to see from Earth, Uranus has two sets of faint rings surrounding it. Some of the larger rings are surrounded by belts of fine dust.

## ICE, ICE BABY

At times, Uranus is the coldest planet in the Solar System, plummeting to minimums of -224 °C. The gas giant's core is made of rock and ice. Water, ammonia and methane ice crystals in the upper atmosphere give the planet its pale blue glow.

## CHANGEABLE CLIMATE

Neptune is an extreme weather zone! It experiences similar freezing temperatures to Uranus, and winds tear across its surface at speeds of 1,200 kmph.

## RADIANT

Neptune gives off more heat than it receives. How it manages to do this when the planet is so far away from the Sun remains a mystery.

## DARK TIMES

The storms on Neptune are so big and ferocious, they manifest as dark spots on its surface. In 1989, one of the largest storms ever recorded was observed in the planet's southern atmosphere - the Great Dark Spot raged for around five years.

## CHANGING STATUS

Up until 2006, Pluto was considered to be the ninth planet in the Solar System. It has since been officially reclassified as a 'dwarf planet'. Heated arguments about whether this decision was correct still go on.

## WEIGHT A MINUTE

Pluto's heart-shaped basin (Sputnik Planitia) has a pull so strong and heavy, its mass may have shifted Pluto's axis. The sheet ice plain is nearly 1,000 km wide.

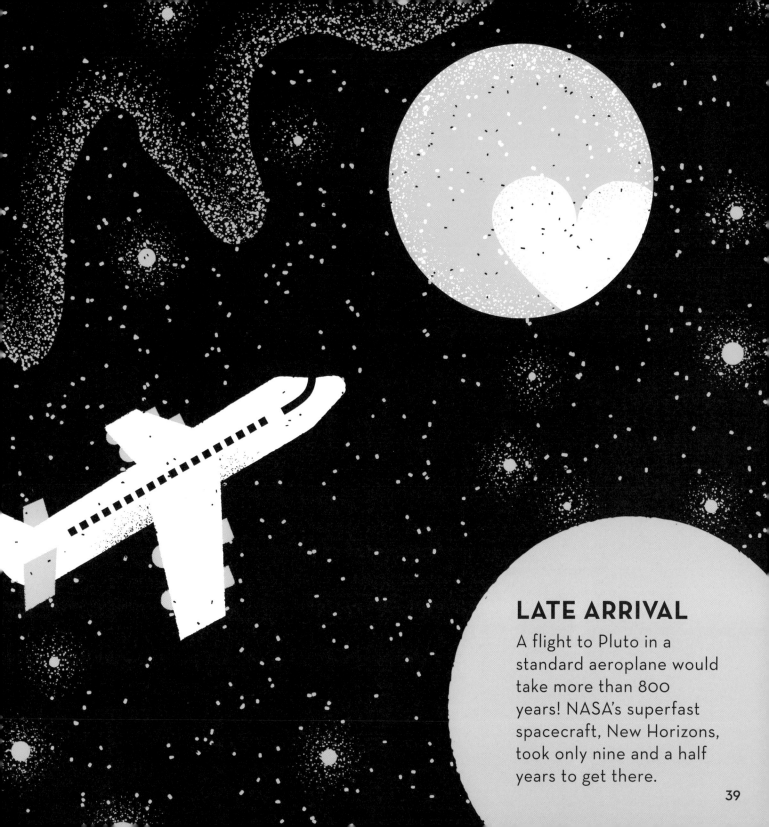

## LATE ARRIVAL

A flight to Pluto in a standard aeroplane would take more than 800 years! NASA's superfast spacecraft, New Horizons, took only nine and a half years to get there.

## MANY MOONS

A moon is a natural satellite that makes an orbit around a planet. Earth has one, but there are many other moons in our Solar System.

## MINI MOON

Some larger asteroids have their own moons.

## LONELY PLANETS

Neither Mercury nor Venus have rings or moons.

## MEGA-MOON

Jupiter's moon, Ganymede, is the largest in the Solar System. It is only slightly smaller than Mars.

## LAVA FOUNTAIN

Another of Jupiter's satellites, Io, is known for violent volcanic eruptions.

## NAME THAT MOON

Saturn has 53 moons that have been named, and another 9 that are still waiting to be given an official title.

## TWO TONE

One of Saturn's moons, Iapetus, has a peculiar two-colour surface.

## LOCAL HOTSPOT

To prove that it is possible to beam an internet signal across a great distance, researchers at MIT (Massachusetts Institute of Technology)and NASA have created a WiFi hotspot on the Moon! Four separate telescopes on Earth transmit the signal to a receiver mounted on an orbiting satellite.

# MOONQUAKE

The Moon experiences quakes that are similar to ours on Earth. This seismic activity can be provoked by many different factors including the impact of small meteors and the tidal pull of the Earth.

## MOON DANCE

Scientists believe that any footprints left on the Earth's Moon could stay there, undisturbed, forever. No atmosphere means no wind erosion; and no liquid water means no water erosion.

## RED PLANET

When there's a total lunar eclipse, the Moon can take on a reddish tinge. Known as a 'Blood Moon', the effect occurs as the Moon passes between the Earth and the Sun, blocking light from the Sun, which makes the Moon appear red.

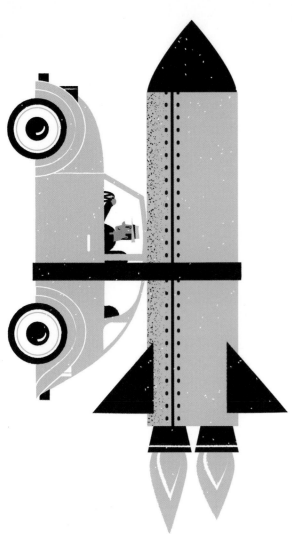

## STRAIGHT TO THE MOON

Outer space begins at the Kármán Line. If you could drive a car vertically upwards at 95 kmph you would be in space in under an hour.

## JUST AN ILLUSION

When in a car the Moon appears to follow you, even though objects on the horizon are moving. This illusion is the result of the distance of the Moon from Earth (384,400 kilometres), which is very far compared with objects here on Earth.

Not only this, but because of the curvature of the Earth you are not moving nearer or further away from the Moon. You are always the same distance from it.

## STACKS ON STACKS

If all the plastic building bricks ever made were stacked together you could make a tower that was ten times as tall as the distance to the Moon.

# SPEEDY STAR

A runaway star moves through space at abnormally high speed. Scientists are able to track the glowing trail of particles the star picks up as it goes. A star becomes a runaway when the gravity of other stars within a system forces it into motion.

## SHADY SUN

Strange, fluctuating light coming from star KIC 8462852 had some astronomers wondering if the view was being blocked by alien megastructures. The real cause of this odd behaviour was actually light filtering through space dust.

# SPACE MOUNTAIN

The brightest asteroid in the sky, Vesta, has a massive mountain near its southern pole. It is even taller than Mauna Kea in Hawaii, Earth's biggest mountain when measured from the ocean floor.

## ICE TO MEET YOU

The centre of a comet is made of ice. Its average size is usually less than 10 km across.

## TAIL OF TWO

Comets have two tails – a dust tail and an ion tail created by ultraviolet sunlight. Both types of tail can stretch out for millions of kilometres into space.

## ANCIENT REMAINS

Made mostly of sand, ice and carbon dioxide, comets are the leftovers from the creation of our solar system.

## HEAVY METAL!

Asteroids are rich in precious metals as well as water. It is likely that in the future gold, silver, titanium and other metals will be mined from asteroids for construction and other projects.

## ROCK ON!

Some asteroids are blown-out comets. The ice on them has gone, leaving just rocky material behind.

## SPACE DRIVE

About once every year, an asteroid the size of a car enters Earth's atmosphere. Luckily it burns up before reaching us.

## CLOUDY AND WET

There are large reservoirs of unfrozen water
floating through the Universe. The largest water
cloud found to date contained 140 trillion times
the mass of water in the Earth's oceans.

# GEM OF A PLANET

55 Cancri e is a planet in the Milky Way known as a Super-Earth. At least a third of the exoplanet's mass is likely to be made of pure diamond. This planet is around 40 light years away, in the constellation of Cancer, so it is unlikely that we will be mining it anytime soon.

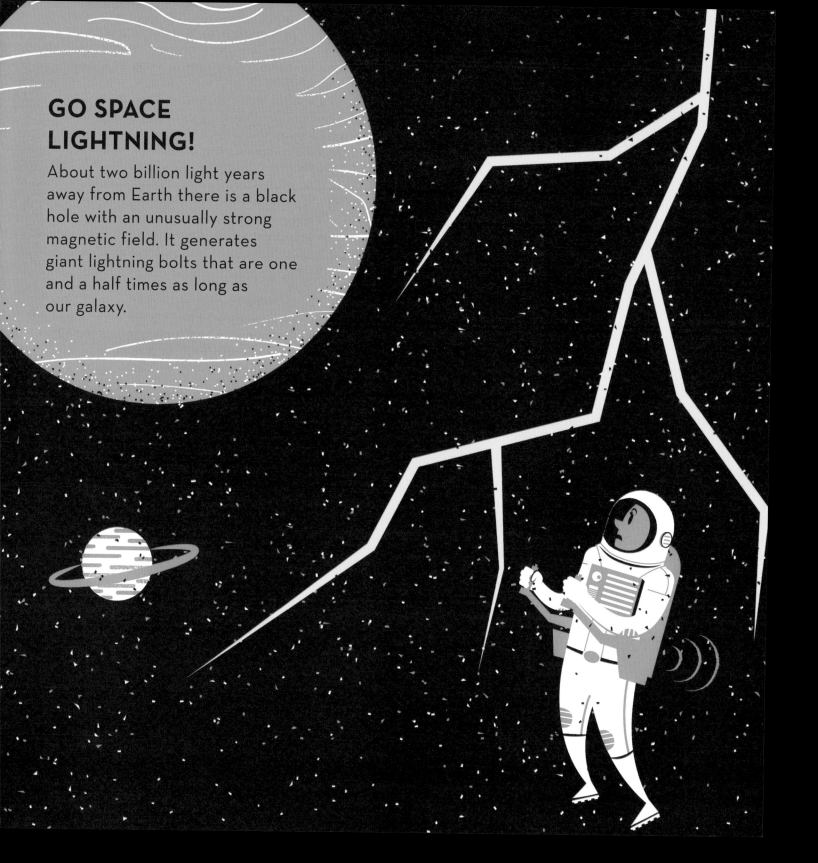

# GO SPACE LIGHTNING!

About two billion light years away from Earth there is a black hole with an unusually strong magnetic field. It generates giant lightning bolts that are one and a half times as long as our galaxy.

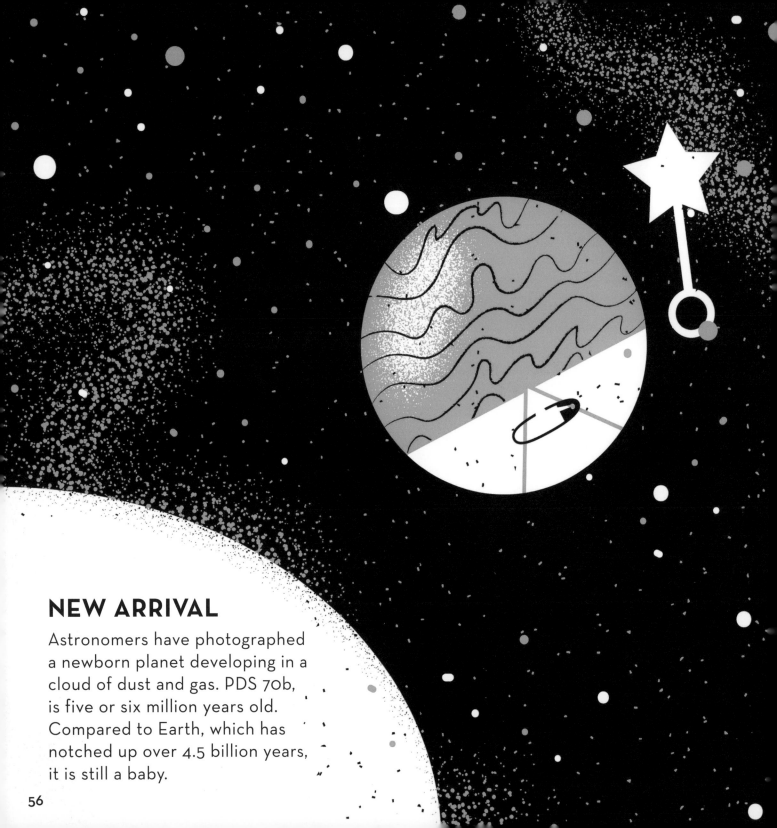

## NEW ARRIVAL

Astronomers have photographed
a newborn planet developing in a
cloud of dust and gas. PDS 70b,
is five or six million years old.
Compared to Earth, which has
notched up over 4.5 billion years,
it is still a baby.

# SLIMED!

Scientists believe that space could be a little sticky. Toxic grease has been discovered floating through the Milky Way. The oily goo is a type of hydrogen-bound carbon blasted off stars.

## COSMIC CLOSE-UP

Gravitational lensing takes place when the gravity of a massive object in space, such as a galaxy or a black hole, bends the light around it. This bending causes a magnifying effect that an orbiting telescope can photograph.

# RIPPLES IN SPACE-TIME

Over a hundred years ago, Albert Einstein predicted the existence of gravitational waves. In 2015, he was proved right. The waves are created when objects move at high speed. Although smaller than an atom, they help scientists to study how the Universe was created.

## CATCH SOME RAYS

In space, astronauts sometimes see random flashes of colour caused by cosmic rays hitting their optic nerve. The Earth's magnetic field protects us from these rays here on Earth.

## STAY FOCUSED

The largest photo taken to date is of
a portion of our nearest neighbour,
the Andromeda 2 galaxy. It's a huge 1.5
billion pixels. The average picture on the
Internet is 72 pixels per inch.

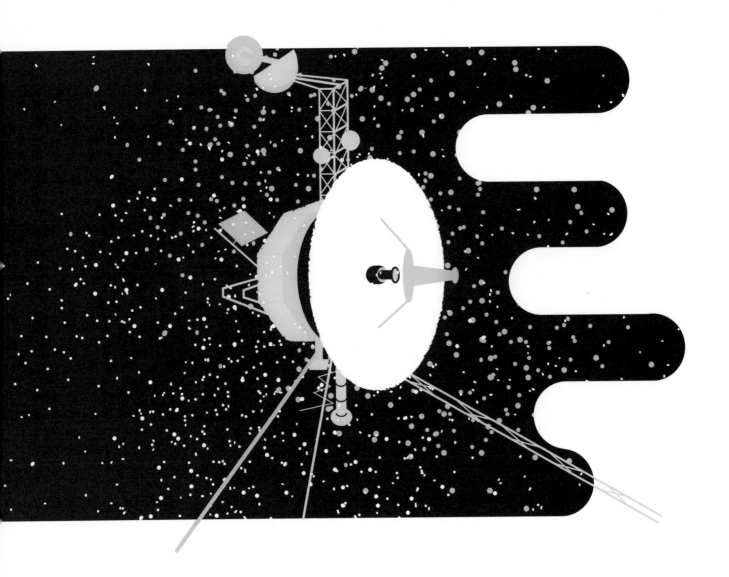

## BON VOYAGE

Voyager 1 is the furthest man-made object from Earth. It is currently over 21.2 billion km away – 16 billion km past the orbit of Pluto. Voyager carries a golden record on board – it will only be played if the craft encounters advanced alien life.

## SPACE JUNK

After years of launching spacecraft and satellites into space, there are more than 500,000 pieces of debris circling the Earth. Scientists are working on plans to try and clear up the planet's orbit.

# RISE AND SHINE, REPEAT

Astronauts living on the International Space Station (ISS) get to see a sunrise every 90 minutes! The ISS moves at a speedy 27,500 kmph, orbiting the Earth every hour and a half.

# TEAR DROP?

It's impossible for a person to cry on a space mission. Astronauts' eyes can well up, but due to zero-gravity the tears will never fall from their eyes.

## BUBBLE TROUBLE

If astronauts on the ISS don't have a good airflow around them when they sleep, a bubble of their own exhaled carbon dioxide can form around their heads. This can make their sleep restless – with possible breathing difficulties and headaches.

# SPACE BLOOMS

In November 2015, NASA astronauts began growing zinnia seeds on board the ISS. By January 2016 the flowers were in full bloom. The crop is part of a research program looking at ways of growing food for long-range space missions.

## FOOT WORK

In space, the rough skin on your feet peels off! Astronauts don't need to walk around under the force of gravity so the soles of their feet become soft and pink like a baby's.

## BATHROOM BREAK

Low gravity makes it difficult to tell if you need to use the loo! Astronauts are trained to go to the bathroom every two hours, because they can't gauge naturally when their bladders are full.

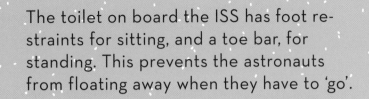

The toilet on board the ISS has foot restraints for sitting, and a toe bar, for standing. This prevents the astronauts from floating away when they have to 'go'.

## IT'LL COME BACK TO YOU.

An astronaut threw a boomerang in the ISS and it returned to him. As long as there is air to provide the necessary forces, a boomerang will return to its thrower – even while weightless.

# RING REDISCOVERED

US astronaut Ken Mattingly lost his wedding ring on a moon mission in 1972. Luckily it was found again nine days later during a spacewalk!

Other items lost in space include a tool bag worth $100,000, a camera and a spatula. Items sent into space include Luke Skywalker's Lightsaber (a prop from 1983), Lego, a Buzz Lightyear toy, dinosaur bones and lots of music – notably by The Beatles, Blur and Louis Armstrong.

# METAL BAND

In space, due to a lack of oxygen it is possible for metal to cold weld and stick together. On Earth, the oxygen in our atmosphere forms a thin layer over the metal that prevents this from happening.

# UNDER PRESSURE

Astronauts train for two years before they launch
into outer space. To prepare for the weightless
environment, they practise spacewalks underwater.

# SPACED OUT

The average NASA spacesuit costs almost 12 million dollars. It is designed to protect the wearer against all of the dangers of space.

## TAKE A WALK

Spacewalks can last around six and a half hours. The life support system inside the astronaut's suit provides oxygen and blocks out harmful radiation.

## MICRO GUARDS

A spacesuit has seven layers of shielding to protect it from sharp meteorite debris.

# GRAB A DRINK

Astronauts wear a drink bag inside their spacesuit. It has a special straw that they bite to get water when they feel thirsty.

## SCRATCHING POST

Astronauts can't touch their face during a spacewalk! A patch of material is fitted inside their helmet instead, allowing them to scratch their noses.

# SMELL OF SUCCESS

NASA employs one individual to smell objects before they go into space. Smells are not easily eliminated on the ISS! One person carefully smells everything to make sure it won't gross out the astronauts. Items that have failed include camera film, felt-tipped markers, mascara and even stuffed animals!

## WASH AND GO

How do you do your laundry when the nearest washing machine is 400 km away? Astronauts wear their clothes as much as they can, and then they are discarded. There are different uses for these garments – some have even been used to grow plants on the ISS.

# GET MEOW-T OF HERE

While over 500 humans have visited space, many animals have also made the journey. Félicette of France was the first and only cat to go on a mission. She reached an altitude of 160 km before safely returning to Earth.

## DOG-GONE

The record for the longest canine space flight belongs to Russian hounds Veterok and Ugolyok. They orbited the Earth for 22 days before coming home.

## SPACE SPIDERS

When a high school student asked if spiders could make webs in zero-gravity, NASA decided to launch their first arachnids into space. Crowned orb-weavers, Arabella and Anita, spun webs aboard Skylab 3 for 59 days.

First published in the United Kingdom in 2019 by
Pavilion Children's Books
43 Great Ormond Street
London WC1N 3HZ

An imprint of Pavilion Books Limited.

Publisher: Neil Dunnicliffe
Editor: Mandy Archer
Assistant Editor: Harriet Grylls
Art Director: James Olstein

Text and Illustrations © James Olstein 2019

The moral rights of the author and illustrator have been asserted.

ISBN: 9781843654032

A CIP catalogue record for this book is available from the British Library.

10 9 8 7 6 5 4 3 2 1

Reproduction by Mission Productions Ltd, Hong Kong

Printed by 1010 Printing International Ltd, China

This book can be ordered directly from the publisher online at www.pavilionbooks.com, or try your local bookshop.